09.18

KT-222-952

F**K PLASTIC

F**K PLASTIC

101

ways to free
yourself from plastic
and save the world

SEVEN DIALS

First published in Great Britain in 2018 by Seven Dials
An imprint of Orion Publishing Group Ltd
Carmelite House, 50 Victoria Embankment, London, EC4Y 0DZ

An Hachette UK Company

1 3 5 7 9 10 8 6 4 2

Text © The Orion Publishing Group Ltd 2018

Hardback ISBN: 9781841883144
Ebook ISBN: 9781841883151

CONTENTS

INTRODUCTION

Plastic: what's the big deal?

Plastic still remains a pretty great invention – syringes, hip replacements, protective helmets, your laptop, my phone, that car. Let's be honest – plastic ain't going nowhere. But that's the problem in a nutshell – all the single-use plastics we buy each day without realising ain't going nowhere either. A plastic carrier bag is used on average for 12 minutes[1] – but it'll still be here in 100–300 years. The water bottle you picked up at lunch could still be here in 450.

Most plastic is recyclable, but lots of recycling plants can't keep up with the amount we consume each year and sometimes the amount of energy required to recycle a product makes the exercise futile. This means a lot of plastic that goes in the recycling bin, along with the plastic put in general waste bins, will end up on a landfill site. This is obviously not ideal (cue scenes from *Wall-E*) – but a whopping 8 million pieces of plastic also enter our oceans every single year.[2] A third of that is from items dropped from ships or lost at sea; the remaining 5-and-a-bit million pieces come from rubbish left on the beach; litter that's dropped in towns and cities that ends up in rivers and drains; industrial spills; badly managed landfill sites and bins near the coast; and stuff flushed down the loo. This all amounts to the following number of microscopic pieces of plastic currently existing the sea:

51 trillion

That's a pretty huge number to get your head round. It's 500 times the amount of stars in the galaxy.[3] And it's a problem for many reasons. One is that none of us wants to be wading through muck when we swim in the sea or spend a nice day on the beach. But more importantly it can be deadly for wildlife – fish, dolphins, seabirds and seals can either become entangled in it or ingest it. Some experts estimate that by 2050, 99 per cent of seabirds will have plastic in their stomachs.[4]

It affects us more directly, too. Over 30 per cent of the fish caught for our dinner tables have ingested plastic – so if you eat fish, you could be ingesting it as well.[5] Scientists are still researching how bad this is for us, but as plastic absorbs chemicals in seawater which have been linked to endocrine (hormone) disruption and even some cancers, it's a pretty bleak reality.[6]

Can we clean up the plastic in the sea?

Only one per cent of plastic pollution floats, and the vast majority of it becomes microscopic. So even if we somehow managed to get all the countries in the world to fund a clean-up of the seas, it would be near-on impossible to actually achieve – and even then if we could, where would we put it? That 8 million tonnes of plastic we unwittingly pump into the sea every year is a lot of rubbish to find a three-century home for.

So what's to be done?

I'm glad you asked! What we need to do is put a stop to plastic pollution now so that the damage doesn't get any worse. In the last couple of years, the conversation around plastic has been growing louder and louder – we need to keep that noise going until all single-use plastic is banned for good and alternatives are found. There have been some brilliant initiatives happening around the world that show we're all heading in that direction. Here are just a few . . .

🖐 Many countries are introducing bans on plastic: France banned plastic cups, plates and cutlery in 2016; Scotland has banned plastic cotton buds; Karnataka in India has banned the use of plastic across the state; and some countries, such as Taiwan, have even banned all single-use plastics altogether.

🗣 Lots of places, including Morocco, Tasmania, France and some states in the US have either banned plastic bags or introduced a tax on them. The plastic bag tax in the UK meant it produced 6 billion fewer plastic bags in the first year it came into force – an 83 per cent reduction rate.

🗣 Many corporations have vowed to go single-use plastic-free – the BBC plan to do so by 2020 and single-use plastics are now banned within UK Government departments. Brands are jumping on board too. Evian has pledged to use recycled bottles by 2025 and Coca-Cola is going to aim to collect and recycle all its packaging by 2030.

🗣 Supermarkets are joining the conversation. The Amsterdam branch of the Dutch supermarket chain Ekoplaza was the first to set up a plastic-free aisle; and Iceland was the first UK supermarket to pledge to be plastic-free for its own-brand products by 2023, with others declaring similar moves.

🤛 Australia, Germany, Finland and many other countries have adopted a deposit return scheme for plastic bottles, meaning you pay a small fee for a bottle which is refunded when you return it.

🤛 Scientists are looking for ways to make a biodegradable alternative to plastic that will give us a material with the same durability plastic provides, but won't harm our oceans or stick around for centuries. One solution being developed uses casein, a protein found in milk, to make a material similar to polystyrene that can be used for packaging.

And there's lots more that YOU can do too:

🤛 **Go litter-picking.** If you live near a beach, then get the community spirit going by becoming part of a beach clean – you can find one near you or get advice on how to organise your own from one of the many beach clean-up charities that exist. If you don't live near a beach, don't let that stop you! If you see litter on the side of the road, bin it – don't let it go down the drain or be washed into a river.

🤛 **Use your voice.** Let businesses know that you've found their waste in places it shouldn't be. If you find a plastic bottle on the beach, take a photo and send it to the company on social media with the location and the hashtag #ReturnToOffender.[7] And if you can afford it, send it to their Freepost address with a note afterwards!

🤜 **Lobby your MP.** Governments are finally taking notice of the plastic issue now – but don't let them forget! Tell them to keep it at the top of the agenda by tweeting or emailing them.

🤜 **And the easiest thing of all? Read this book!** There are 101 ideas in here to help you, your friends, your family, your neighbours, the person sitting next to you on the bus and everyone else you meet along the way to change the habits we've all grown up with – until we no longer rely on plastic in the way we currently do. So give them a go and spread the conversation as far and wide as you can – in person and online with #fkplastic.

Together we CAN do it – and we will.

FOOD

AND

DRINK

#1 Buy fresh over frozen

Frozen is obviously great when it comes to prolonging the shelf-life of food that would otherwise go bad quickly and have to be thrown out – waste in itself. But the majority of frozen veg and fruit comes in plastic packaging. However, you can still freeze your food – instead just buy fresh and freeze in reusable containers when you get home. Sorted!

#2 Beeswax food wrap

Who wants clingfilm anyway? We've all been there trying to unstick it when it's rolled in on itself. So here is an alternative for you – beeswax food wrap. It might sound a bit hippy, but not only are these wraps biodegradable, they're also reusable for up to a year.

bees
against
plastic

#3 Source your own sauce

Who here likes sushi? Who here likes soy sauce? *Me me me* But if you're getting it in a takeaway shop, chances are the soy sauce will come in a teenie little plastic bottle. Avoid that bottle! Buy your own bottle of soy sauce for your kitchen cupboard/to keep on your desk at work/in your bag if you're a real addict. The bigger you buy, the less packaging you use.

#4 Invest in lunchtime

The nicer your lunchbox, the more likely you'll want to use it/remember to pick it up after you left it to dry on the rack at work. FACT. So treat yourself to a nice steel tin with your initials engraved on it (also makes a good plastic-free gift for fellow like-minded individuals).

#5 Bring your own bottle

You know this one already – but it's amazing how many plastic bottles of water and other drinks are still made and bought each year; it's currently estimated that a million plastic bottles are bought around the world every minute.[8] It's not that plastic bottles aren't recyclable – the majority are – but with staggering numbers like that, recycling efforts just can't keep up. The answer is simple: buy yourself a nice, snazzy reusable bottle, keep it in your bag at all times and top it up at water fountains on the go.

#6 Opt for the ice cream cone

Everyone knows one. That person who instead of getting their ice cream in a cornet asks for the scoops to go in the little tub with the little plastic spoon. We know, we know, some people don't like the wafer (if you're one of them then you're a little weird, but we're not here to judge). But even if you DON'T like the cone and you've somehow ended up surrounded by other people who won't finish it off for you either, as much as food waste is a plague in itself, we know which container is biodegradable. (Ice cream for president.)

#7 Be choosy about cheese

For a lot of us, cheese can be a big deal. But how to escape the fact it often comes wrapped in plastic? Here are our top tips: first, see if you have a local cheesemaker nearby you can try – the likelihood will be you'll not only be avoiding plastic packaging by buying from them, you could also be reducing the mileage between you and the source of the product (and bonus points for supporting local businesses). Second, if you don't have a local cheesemaker/you don't love cheese that much to warrant the price, take a container to the cheese counter in the supermarket and ask them to pop your chosen cheese in it. Third, some cheeses do actually come in plastic-free packaging: truckles of cheese are a good example (minus any plastic stickers), and some brands go for cardboard boxes (just beware of plastic wrapping inside them).

#8 You deserve your own goblet

Takeaway coffee cups are for the average (cup o') Joe – you want one that shows your personality and doesn't give away that you support huge conglomerates. PLUS a lot of coffee shops will actually give you money off your hot drink if you bring your own reusable cup.

P.S. Some coffee shops (and countries) are already wise to the plastic pandemic and are using biodegradable cups made out of recycled materials 🐨. But it's still obviously better to avoid one-off cups wherever you can.

P.P.S. If you do find yourself in desperate need of a hot beverage and you've accidentally left your reusable cup on the draining board, forego the plastic lid and walk more carefully. Every little helps.

#9 Shop at the bakery/green-grocer/fishmongers/butchers

Chefs love telling us to do this but is it really practical? It might not be as convenient as getting all your food from one place, but on the major plus side it can often be cheaper and the food can be better, AND you get to feel warm inside for supporting small businesses. Obviously these shops still do use plastic – so make sure to take along bags and containers and ask the person serving you to refrain from using plastic packaging. Alternatively, you can ask the counters in supermarkets if they'd be happy to do that for you too.

#10 Pack a bag

Lots of countries already tax plastic bags and boy has it made a difference to how many are produced each year. But we can all cut back more. Plastic bags are still available and it's all too easy to pop to the shops after work for a couple of necessities and find yourself empty-handed. The solution is those very sexy fold-up shoppers that come in a pouch – they fold up small so you can keep them in your handbag/rucksack/shoulder bag/donkey saddle and whip them out whenever you need them. Lots of department stores sell them but you can also track them down ever-so-easily online by searching 'fold-up shopper'.

#11 Shop at zero-waste supermarkets

Not everyone has a zero-waste grocery store near them, but more and more are popping up so keep your eyes peeled! Zero-waste supermarkets are shops where you can collect your dried goods in your own containers, thereby foregoing the plastic packaging (or any other packaging for that matter) that rice, pasta and the like are usually to be found in. Manchester United fans – get down to Devon, England and visit Earth. Food. Love, which is run by former Man U footballer Richard Eckersley and his wife Nicola.

#12 Look for plastic-free aisles

Coming to a supermarket near you! Some supermarkets around the world are planning an aisle of their own products which will be entirely plastic-free. Ekoplazer in Amsterdam was the first in the world to get one up and running, whilst the big chain Iceland in the UK was the first to pledge to go completely plastic-free on its own brands by 2023.

#13 Grow your own herbs

For herbs such as basil, mint, coriander and parsley, growing your own is SO easy, we promise. And it saves you having to buy either plastic-packaged fresh herbs or those little plastic pots of dried herbs, which will also be cheaper for you in the long run. So long as you have a part-sunny garden/balcony/windowsill, a pot, some compost, and a good enough memory to water the thing, you can do it. You don't even need to head to a garden centre anymore to buy the seedlings (baby plants) – most supermarkets sell them nowadays and, if you look after them, they'll give you a return for months and months.

#14 Grow your own salad

Whilst you've got your trowel in your hand re the herbs (kidding, you don't need one), spare a thought for the next absolutely super-duper-easy-peasy foodstuff to grow: salad leaves. A tray, some compost, some seeds, some water. Place the tray somewhere sunny, inside or out. You should see some little shoots appearing within two weeks, and once the leaves are roughly 8cm/3 inches, you can cut them with scissors whenever you'd like some. Like a Christmas miracle, the leaves will grow back and you can keep cutting to your heart's content (this does stop eventually – it's not really a magic trick. Then all you need to do is replace the compost and re-sow the seeds). The upside to only 'harvesting' when you want to eat the salad leaves is no soggy gone-off lettuce at the bottom of your salad drawer, and ultimately more money in your pocket as a result.

#15 Swap crisps for doughnuts

Too right we're serious! Sure we all know avoiding both the doughnut and the crisps would be better for our health (pft), but if you're going to reach for a treat anyway, make it a loose baked product like a doughnut or cookie over a pack of crisps and biscuits. Many of the latter are packaged using a metallised plastic film, which theoretically could be recycled but a lot of the time isn't due to the cost. Loose baked products on the other hand are totally fair game 👍.

#16 Switch to loose tea

Some mainstream teabags use polypropylene – a type of plastic – to seal the bags together. This is a bit of a bother considering in the UK alone, 165 million cups of tea are drunk each day and 96 per cent of those are estimated to be made using teabags.[9] On top of this, you might have noticed that the cardboard box containing teabags is sometimes wrapped in cellophane. Many well-known teabag brands are working on removing plastic from their products, but in the meantime, look out for brands that don't use polypropylene, or switch to loose tea, avoid the plastic packaging and feel fancy at the same time.

#17 Pick loose fruit and veg

Don't bother with the avocados that come two in a pack, or the bell peppers that come in threes, or the shrink-wrapped broccoli. Especially don't bother with the half portions of cucumber you now find in supermarkets which come shrink-wrapped and *then* packaged in another layer of plastic. Opt instead for the veg that's loose in trays, and – if you can – also buy from places that shun sticky labels.

#18 Bring fabric bags for fruit and veg

Don't be tempted to bag up your loose fruit and veg with the plastic bags provided. We surely all have a backlog of tote bags by now, right? If you're buying a load of potatoes, Brussels sprouts, apples or the like, then take those (clean!) tote bags with you so you can load up your loose veg and fruit and easily weigh them when it gets to checkout time.

#19 Lunch club

If you work or study in a town or a city, you'll be all too familiar with the temptation of popping out for lunch instead of making your own the night before. And yes, it requires a little effort, but not only does it help you avoid plastic-packed sandwiches and salads, it tends to be cheaper too. Some estimates suggest that city-workers spend £2,500/US$3,337 a year on lunch![10] And who really likes dry crusts and overpriced lettuce anyway? Invest in a packed-lunch recipe book to motivate you, or google 'best packed lunch ideas' for inspiration.

#20 Cut your own fruit

Introducing: the knife. We know – mind-blowing. Melon, mango, pineapple – all of them and more besides now come chopped up in plastic pots topped with plastic film. We know it's convenient to buy them pre-prepared, but if you buy the fruit in its original, natural packaging you can chop it yourself without any of that plastic and get to feel gloriously smug.

#21 Make juice not war

It's not just the lid on juice cartons that's plastic; many cartons are made up of 20 per cent polyethylene.[11] That doesn't mean it's not recyclable – it is – but just like plastic bottles, there often isn't the resources or capabilities available to recycle all of them. But if you're a real juice fan, bypass the preservatives and added sugar that might be hiding in shop-bought juice, invest in a juicer and get squeezing those oranges!

#22 Eat out, not in

If your favourite takeaway comes in plastic tubs, snugly packed shoulder to shoulder in a plastic bag, with plastic pots of sauce on the side, it may be time to think about getting off the sofa and treating yourself to sitting in an actual restaurant instead (or you could make your own dinner – but let's face it, there are times when we all just can't be arsed). If you pick up your takeaways instead of getting delivery, you could instead ask the restaurant while ordering if they'd be happy to package your requests up in the containers you're going to bring.

#23 Bring back the milkman

In 2016, only three per cent of milk in the UK was put in glass bottles, but as retro as it may sound, it's now on the rise! Reportedly 200,000 more glass bottles of milk are being delivered each day than they were a couple of years ago.[12] If you live in the UK you can search for your nearest delivery services using findmeamilkman.net. If you live outside it a good google should turn an alternative service up. It's more expensive than what you get in the supermarket, but you're getting the convenience of delivery and it's still a dying trade compared to the supermarket, so who knows – if the support for milkfloats grows enough, competition might soon drive those prices down.

hey there!

#24 Jarred deli foods

Don't you just love olives? And chargrilled artichokes? Mezze/nibbly bits are awesome – just don't go looking for them in the fridge aisles of the supermarket as they nearly always come in plastic tubs. You want to hop on over to the jarred goodies – lots of the same stuff, they're often cheaper and they last for longer.

#25 Make your own

Did you know it's super-easy to make your own bottle of ketchup? It's just as tasty and you'll probably end up with something that contains less sugar and salt. Instead of buying the store-bought pots of hummus and tzatziki, you can also give them a whirl at home – we guarantee they will taste better and they only take 15 minutes to whip up. Easier swaps still are the sandwich fillings that come ready-made like egg mayonnaise – you can boil and peel that egg and mash it with mayonnaise, we believe in you!

#26 Say goodbye to gum

Have a guess how many pieces of gum are made in the world each year.

If you happened to say, '1.74 trillion',[13] give yourself a pat on the back. Now have a guess what most chewing gum is primarily made from. That's right: a type of plastic. Pass us the mints.

#27 Buy the can, not the bottle

The best thing to do would be not to buy either a can or a plastic bottle as both materials are a drain on energy and natural resources. What you want to be doing is keeping on using your old faithful: the reusable bottle. But if you're desperate for a drink and there's no water fountain around, go for the can. As a planet, aluminium is the material we recycle the most, to the point that 70 per cent of cans are now made from recycled aluminium. Each one of these requires only 8 per cent of the energy needed to make a brand-new can from scratch, and if they do find their way on to landfill rather than the recycling bin, they don't transfer any nasty chemicals into the ground.[14] Plastic bottles on the other hand DO transfer nasty chemicals and only half of them end up being recycled.[15]

#28 Don't go for the multipacks

It might cost you a tiny bit extra, but don't bother with the packs of tinned tuna or cans of pop that are wrapped together with plastic. Those tins sitting lonely on the shelves are just crying out to be picked – make their day.

pick me!

#29 Carry cutlery

We don't mean every time you sit down at a restaurant or pop over to your mate's house that you need to whap out your own knife and fork from your inside coat pocket. That will probably come across as a bit rude. This is more if you've given in to buying lunch on the move, or if you're eating in a cafeteria; if you carry your own knife, fork and spoon you don't need the disposable ones. Result.

#30 Charcoal your water

If you like to filter your tap water, then swap your plastic filter cartridges for charcoal, which has been used by us humans to filter water for thousands of years. It won't remove everything leading brands of plastic water filters claim to do, but it will dispense of the chlorine, sediment and volatile organic compounds you might not want in there.[16] It's also more cost effective as the filtration properties of charcoal lasts for longer. We recommend Black + Blum Eau carafe, but you can find various options online.

#31 Say 'no' to straws

Many countries are doing this on your behalf, but if you're in a locale where supermarkets, restaurants and bars etc. are still sporting the plastic straw, just say 'no'. If you really like drinking through a straw or you have a specific reason why you need to, you can source metal or glass ones on the internet and in some high street shops (steer clear of paper ones as these are still disposable). If you have a medical condition that requires a 'bendy' straw then biodegradable options are available online e.g. from Plastico.

#32 Think coffee

Most home coffee machines involve plastic parts or packaging in some way, but you *can* reduce how much of it you're throwing in the bin. Machines which use plastic capsules aren't great, whereas ground coffee is compostable, so if you're looking for a coffee maker think French press; percolator (ideally stainless steel); or drip coffee maker.

AROUND

THE

HOUSE

#33 DIY cleaning products

There are many reasons to give DIY cleaning products a go: they often use products you have lying around the house or which are cheaper than shop-bought cleaning products; they're better for the environment and arguably our bodies than using harsh chemicals; and of course, by not buying plastic-bottled disinfectant and the like, you're cutting down on packaging. Did you know that diluted white vinegar can be used as a bathroom cleaner and floor cleaner; undiluted white vinegar can be used for the toilet and showerhead; and baking soda can be used to tackle grime? There are several recipes and blogs online dedicated to how to use these – give them a go today.

#34 Cardboard over plastic

If making your own cleaning products isn't something that floats your boat, then see if you can swap any of your regular plastic-packed products for cardboard packaged ones instead, especially ones where the cardboard is made from recycled materials. Also keep your eyes out for changes cleaning product manufacturers make to their packaging: Ecover is making waves in this area, including launching a bottle made from 50 per cent recycled plastic and 50 per cent plastic found from the sea (their aim is to use 100 per cent recycled plastic to make all their bottles by 2020), and Method products also use 100 per cent recycled plastic to make their bottles.

#35 Buy big

If you do have to buy a plastic-packaged product for whatever reason — be it washing up liquid or shampoo — buy big. You're going to get through it all eventually, it's usually more cost-effective to do it that way, and there's a smaller surface area in buying one big bottle than two smaller bottles, which means less plastic overall.

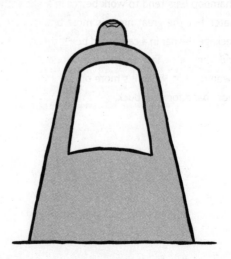

#36 Solid shampoo and conditioner

Type 'plastic-free shampoo' into your search engine and you'll find lots of alternatives to the big bottles currently residing in bathrooms the world over. The majority come in a solid bar and can be found online as well as in some cosmetic high street shops. Shampoo bars tend to work better in areas with soft water, but the great news is most brands purposely package the bar in a cardboard box rather than plastic and they also last longer than squeezy shampoo, meaning that, while it's more outlay initially, you get more bang for your buck.

#37 The bar vs the dispenser

A super simple one for you – stop buying a soap pump dispenser, start buying the soap bars. Means you can splash out on a nice soap dish and everything.

#38 Detergent grows on trees

Who here's heard of soapnuts? Most people haven't, but they've actually been used to wash clothes for thousands of years. Soapnuts are a rather miraculous dried fruit shell that contains a natural washing detergent called saponin, which is activated when placed in water. You can buy them online and safely use them in your washing machine (pop them in a muslin bag first), and as well as cutting back on the plastic packaging of most types of washing detergent, they're vegan and hypoallergenic.

#39 Swap your toothbrush

Next time you need a new toothbrush, have a search online for ones made of bamboo – they're just as cheap as their plastic counterparts and don't come in any plastic packaging (true of most brands – check this before you buy). Make sure you check the material of the bristles too to make sure they're also plastic-free and biodegradable.

#40 The toothpaste solution

So many toothpastes come in plastic packaging, so what to do? There are lots of homemade recipes for toothpaste powders you can give a whirl, but if that's all a bit out there for you then you can find jarred toothpaste online such as Georganics, who also sell a soap stick that doubles up as a toothbrush, lasts up to six months and comes in recyclable and compostable packaging.

#41 Plastic-free flossing

If you're a keen flosser, then you might be worrying about what to do regarding the plastic pots dental floss comes in – as well as the fact lots of floss is made from nylon. Here's where silk floss comes in – you can find brands all over the shop online (e.g. Dental Lace) that come in plastic-free, refillable packaging and, depending on which brand you go for, the floss can also be biodegradable and compostable.

#42 Say no to cotton buds

Some countries are beginning to ban the plastic cotton bud – if you're not in one of them then make the choice not to support their manufacture. If you're an avid cotton bud fan, though, you can actually find non-plastic alternatives online and in lots of shops (the Cotton Bud Project provides a list of those available in the UK).

#43 Shower less

We're serious. If you can't give up your plastic-packaged shampoos and shower gels, remember that the less product you use, the less packaging you use. And experts say we really don't need to shower daily anyway; depending on your skin type, once or twice a week is fine (and sometimes better if you have dry skin, as washing rids us of our natural oils).[17] You'll also score bonus points for saving on water and energy, and gain a few more minutes in bed in the mornings. Result.

#44 Smell green

Just like going to the toilet, wearing deodorant is something the majority of people feel they (thankfully) can't scrimp on. But roll-ons and aerosols are mostly made from plastic and can be hard to recycle, which means another big hello to the landfill. But hooray – there are alternatives! Solid deodorants (think a stick a bit like a bar of soap) and powdered deodorants are available in tins and glass jars in several high street shops (such as Lush) and online. If you're feeling really creative you can also google homemade deodorant recipes.

#45 Shaving cream

Like shampoo and conditioner, you can get shaving cream in a soap bar form. If you're feeling fancy, buy one that comes in a wooden bowl (refills are often available for when you run out).

#46 Wise washing

Because so many of our clothes contain plastic, sometimes microfibres from them can escape down the drain of our washing machines. Two tricks to help avoid this: first – fill your washing machine up; it'll create less space for friction between the items, ergo less fibres will be dislodged. Second – use a microfibre bag or ball (you can find these online) to collect the fibres, which you can then pop in the bin rather than allowing to go into the drain.

#47 Natural air fresheners

Swap plastic air fresheners for fresh flowers, incense sticks and candles. You could also give eucalyptus branches a go in the bathroom, as the steam from the shower helps release their fragrance.

#48 Plastic-free moisturiser

Look for sticker-free glass jars of moisturiser and swap to serum that comes in glass bottles. If you cook with coconut oil, don't forget that this can double up as an effective moisturiser too.

#49 Plastic-free make-up

A lot of make-up comes in plastic tubes or pots or requires plastic applicators. More and more companies are wising up to the issue and doing their best to reduce their plastic packaging, but for now keep your eye out for refill make-up (e.g. Mac and Nars sell refill palettes for eyeshadows, blusher etc.) and have a search online for make-up brands who already sell plastic-free make-up or sell them in recycled/recyclable containers. One such company is US-based Urb Apothecary, which offers a label opt-out service. You could also follow plastic-free/zero-waste bloggers who suggest and review products: we love Eco Boost by Kate Arnell.

#50 Alternative make-up wipes

Make-up wipes tend to come in plastic packets, not to mention the fact the wipes are disposable – if you use one a day that's 365 wipes you are sending to landfill every year. Now think about how many of your friends do the same, and your friends' friends and so on and you're quickly looking at a very large number. So switch to something that's reusable instead. Invest in some cotton flannels or – if you need something that's smaller or you don't want to buy a lot of them – cut your flannels into small circles that resemble cotton pad size. Stich a line of thread a few mm inside the ring to stop them from fraying, or if you need them to be sturdier, stitch two flannel circles together. Use with hot water or a cleanser of your choice.

#51 Find a wooden hairbrush

Most of us don't need to buy a new hairbrush very often, but next time you lose yours/break it and need to replace it, consider buying a wooden hairbrush or afro comb instead of the regular plastic ones. Re brushes, ones with real hair for bristles instead of plastic spokes tend to get mixed reviews – you're either a lover or a hater – but you can also get wooden brushes with wooden pins which work well.

#52 Controlling your hair

Everywhere you look is a discarded hair tie or hair pin, except when you need one – then you can't find one for love nor money. Pft. But we should all be making more conscious decisions regarding what we use to keep our hair in place as some hair elastics and scrunchies are made from plastic (and generally are made from a non-recyclable synthetic material), or come in or with some form of plastic packaging. But hooray, there are alternatives! The brand Kooshoo makes hair ties in different colours made from cotton and natural rubber (they also sell hair bands and head

bands) which are available on several sites online including their website. Or you could opt for a long hair pin made from a natural material such as wood; there are YouTube tutorials galore on how to use these to make a very stylish and sturdy up do. Also watch out for hair slides/grips which are sometimes made of plastic or come tipped with a plastic ball — look after the ones you've got and try to not to buy any more in the future.

#53 Reusable nappies

This isn't for the faint-hearted. It involves a LOT of washing (which of course has its own environmental impact). But during the first two years, your baby is estimated to get through over 5,000 nappies. Aside from the environmental impact, this has been shown to cost around £800/US$1070.[18] That's a lot of (plastic-free) toys and days out you could be giving your baby instead. Reusable nappies have also come on leaps and bounds since the ones our grandparents were washing by hand using a tub and mangle. To see what other parents are recommeding, have a look at parent bloggers who are local to you.

#54 Toilet paper without the plastic

There are some things you can live without, but going to the loo isn't one of them. If you want to be super green you can use reusable cloths (!), but realistically most of us want to be using toilet paper. Yet toilet paper rolls often come stacked together in plastic — so keep your eyes peeled for plastic-free packaging. Two to try are Greencane (the cellophane used in their packaging is plant-based, biodegradable and compostable) and Who Gives a Crap (their rolls come wrapped in paper instead of plastic).

#55 Return of the match

You're lighting candles/the barbecue/your wood burner. What do you reach for? The lighter that grazes your thumb? Or the old-school flick and fizzle of striking a match? We won't pretend that matches aren't without their own environmental impact, but FYI the majority of lighters bought today are disposable. According to Bic, they alone produce 6 million every *day*.[19] That's a lot of plastic that needs to eventually end up somewhere.

#56 That's rubbish

There are a couple of things you can do when it comes to lining your bins. Firstly, the ideal would be to stop using bin bags altogether – if you have a good recycling system in your part of town then all your wet smelly items would go in your food waste bin (for which you can buy compostable bin bags made from cellulose) and then your recycling waste and the rest of your rubbish can go in their respective wheelie bins without a bag. If for whatever reason this isn't going to work for you, you can buy bin bags online or in most supermarkets which are made from recycled plastic, which is at least cutting back on some of the plastic being sent to landfill.

#57 Flower power

Flowers in the supermarket often come wrapped in cellophane – go to a florist instead (although you'll want to check how they source the flowers too for the same reason) and only take them in brown paper. Skip the ribbon if it's not fabric. Alternatively, if you're keen to fill your house with blooms, get growing sweet peas this summer. Sweet peas come in multicolours, smell delicious, are easy to look after and grow back like magic once you've cut them.

#58 Rubber gloves

Some cleaning gloves, despite being universally referred to as rubber, are made from plastic or a synthetic rubber called nitrile. What we should really be looking for – so long as none of us have an allergy to the materials – are 100 per cent rubber or latex gloves. We also want something durable – if you buy cheap gloves they are going to get holes in or leak quicker. Casabella is one such brand you can try that ticks these boxes, is available online and some high street stores, and Mila Kunis even sported them in that really brilliant movie, *Jupiter Ascending*. If you *are* allergic to latex then you can buy biodegradable nitrile gloves from GREEN-DEX – although these are currently only available in disposable packs of 100.

#59 Swap scourers for natural cloths

Did you know lots of scourers are made from plastic mesh? Swap them and your plastic brushes for a cloth made from wood fibre instead, which still works and is gentler on your dishes and pans. For tougher stains, you can use natural loofahs such as those made by LoofCo – their washing up pads are biodegradable and compostable.

LIFESTYLE

#60 Give a pre-loved toy

The countries in the world that spend the most on toys each year include Germany, France and the US, with the most spoilt kids to be found in the UK:[20] the average British child under nine receives £350 worth of toys each year.[21] That's a lot of dollar for some moulded plastic, especially considering the apple of your eye will probably have outgrown it in twelve months' time. So save your money and as much stuff going in the bin by buying toys from charity shops – and making sure to donate the ones your kids are bored of too. Plus this means you're giving to charity 😇.

#61 Non-stuff stuff

You're a kind and generous person and you want to buy your mate a gift for their birthday or Christmas. But the plastic-panic sets in – what can you possibly give that avoids plastic in the product and the packaging and doesn't contribute to the £2.6 billion of unwanted presents that clutter homes or reach landfill each year?[22] First of all, chill out – you don't need to take all of this on your shoulders. Second, there are plenty of brilliant presents you can gift that avoid all these things: take your friend for dinner or the theatre; gift an experience; get them a membership or buy them an online magazine subscription.

#62 Say Happy Birthday better

E-cards have yet to fully catch on, but it's worth suggesting to your mates and your family instead of sending paper cards – quite aside from saving paper, card and plastic finishes, it'll save you a lot of cash over the years. But if you're a card fanatic then make sure you're buying them loose – cards sometimes come with their friend, the envelope, inside an unnecessary cellophane pocket. So only go for the ones sitting free and easy, and if you can buy cards made from recycled paper then even better.

#63 That's a wrap

Did you know in the UK, an estimated 108 million rolls of wrapping paper are used at Christmas time *alone*?[23] People often think wrapping paper is as easily recyclable as paper, but lots of it contains plastic: the shiny stuff is often due to a laminate form of plastic, and any glitter or foil tends to be plastic too. The best thing to do would be to wrap the lovely gift you've bought in something the person can use again for someone else – like an off-cut of material or a scarf. If that doesn't float your boat then try reusing the free local paper, recycled wrapping paper or brown paper you've decorated yourself.

#64 Paper tape

Sure you can wrap a present up with string, but that's no fun for Aunty Shirley who likes to cover every inch of the paper in sticky tape to ensure unwrapping it is akin to breaking into Fort Knox. Luckily for Aunty Shirley, there are now several brands of biodegradable paper tape available online (just search 'plastic-free paper tape') that are not only made from recycled materials, the sticky side is made from natural substances too.

#65 Fabric ribbon

Last one in our 'how-to-wrap-the-plastic-free-way' series is about ribbon. You know the shiny stuff that crinkles into circles when you pull the scissor blade down it? That's made using a type of plastic. But if you can't give a gift without sticking a bow on it, you can always opt for fabric ribbon – and the good thing about that is the person can keep it and re-use it on a present for you at a later date.

#66 How to shop

A lot of the clothes we wear nowadays are made from synthetic polyesters which aren't biodegradable, as well as nylon which is a thermoplastic. If you're buying new clothes and plastic is on your mind then go for natural fabrics such as cotton, wool, silk, denim or leather (N.B. that some of your other environmental and ethical principles could be offended with any of these fabrics, so we'd suggest doing your own research around these).

#67 New threads

As a planet, we consume WAY too many clothes: an estimated 80 billion items each year in fact.[24] Not only is a form of plastic (synthetic polyester) a prime material of these items, but the amount of energy and water used and emissions created in the production and transportation of these products is staggering. So next time your jeans get a hole in them, why not give sewing it up a go? Or if you're bored with what's in your wardrobe, arrange a clothes swap with your friends or colleagues.

#68 Long live the High Street!

If you're a fan of online shopping, rediscover the joys of going into the shop for a change. When you buy online, many retailers supply each item in a separate plastic bag, meaning that (although there will still most likely be a plastic packaging footprint in store) you'll be saving on plastic packaging by going into the shop.

#69 Hair removal

Hair removal can be a plastic-y business due to the materials used and the packaging involved. Laser hair removal is probably the best for avoiding this, but even if you buy a home kit it can be a lot of outlay in one go (and who knows, one day you might want that bush back). Instead, for small areas you could give threading a go or sugaring, which you can do yourself at home. For sugar paste recipes and instructions www.wikihow.com is very helpful, as is the blog gippslandunwrapped.com. Failing that, invest in a stainless steel razor if you can (not as expensive as you might think – and again, a good useful plastic-free present!) and definitely steer clear of buying disposable ones.

#70 Don't smoke

Don't worry, we're not going all preachy. By now everybody knows smoking is the cause of 90 per cent of lung cancer after all.[25] (Whoops, sorry – totally unintentional.) No, instead we're looking at the environmental impact. Packs of straight cigarettes come wrapped in cellophane; loose tobacco tends to come packaged in a plastic pouch, but quite aside from that, the filters in cigarettes are mostly made from a type of plastic. This is a problem because, even

though it's recyclable as a material, as a planet we buy 6.5 trillion cigarettes every year[26] and the stubs mostly end up in landfill or washed into the sea where birds and fish mistake them for food. The filters can also break down further into many hundreds of tiny fibres and so can't even be cleared up from beaches.[27] How to get round this? Don't smoke. Simple.

#71 Glitter me up

It was a sad day when festival goers everywhere realised that the shiny pieces they were gluing to their heads and bodies were teenie bits of plastic, and the same goes for all the pieces you used in arts and craft at school too. But luckily biodegradable glitter now exists and can be found online and in some high street shops.

#72 Power to the period

Sanitary towels tend to come in plastic packs; tampons come wrapped in cellophane; and if you use a plastic applicator with your tampons then that's even more plastic packaging heading to landfill. The pads and tampons are also of course waste in themselves. Aside from the effects on the environment, did you know having periods can cost you up to £18,000/US$24,000 over the course of a lifetime?[28] That's a lot of dollar! So what's the solution? You can give menstrual cups a go (there are lots of brands to choose from: Mooncup, Lena Cup and Fleurcup are just a few); you'll be surprised how easy they are to use and they save all the waste and the cumulative spend. If that doesn't float your boat then you could try reusable sanitary towels or the new period pants which have been getting rave reviews called Thinx. If you still find yourself attached to your tampon, then go for the applicator-free ones; every little helps.

#73 Let's talk about sex

The condom is a cracking invention: it not only protects from unwanted pregnancy but plenty of STIs too. If, however, you are in a monogamous long-term heterosexual relationship, both you and your partner are squeaky clean but you don't want to make a baby, you may want to skip on to other methods of contraception. Pills come in plastic sheets. Until this changes, if you fancy some plastic-free(er) options then you could consider the injection or the coil: find out more from your doctor or health advisor.

#74 Travel right

Lots of us make air travel cheaper by only taking hand luggage. This leads to the temptation of 'mini' products of liquids. You know the ones: the cute teenie shampoos, conditioners and so on. If you're using bars of shampoo and conditioner then you can get away with taking these in your hand luggage, but if you've yet to be converted (see page 47), invest instead in your own set of plastic miniatures that you can fill up time and again from bigger bottles. This will be ultimately cheaper for you in the long run. If you're travelling with friends, it might also be worth all clubbing together to get one hold luggage bag for all the toiletries, including your razors (see page 83 — don't be tempted to buy disposables at the airport!).

#75 Wear sunscreen

Most high street sun cream comes in plastic bottles, which is a bit of a bother – and there's also the issue that some of the chemicals within the suncreams are harmful to sealife. In May 2018, Hawaii became the first place to ban sunscreen that contains oxybenzone and octinoxate because they contribute to coral bleaching.[29] Luckily, there are already quite a few sun creams around that don't contain these and also come packaged in plastic-free materials. Lush do a solid sunscreen wash called 'The Sunblock' and you can also try products from Butterbean Organics and Dirty Hippie Cosmetics (among many other options which you can find with a quick search online) which tick these boxes.

#76 After-sun skincare

You over-did it or you skipped over the previous page: it happens. How do you soothe your skin the plastic-free way? You can get aloe vera gel in tins or jars on the internet and you can also make your own calamine lotion using mostly ingredients you'll have in your house (you'll likely only need to buy calamine powder or bentonite clay). Blogs online such as frugallysustainable.com and thesoftlanding.com provide easy recipes to follow. You can also look for products in recycled packaging, for example Lush try hard to use recycled plastic material for their products which come in black pots.

#77 Newspaper supplements

It's Sunday – you've popped to the corner shop for a paper because you want to support print media and you enjoy being 'analogue', and paper's recyclable isn't it. Then you get your paper home, go straight for the supplements and have to tear through a *cellophane bag* to get to them. Why oh why?! Politely tweet/email/write a letter to the newspaper in question asking them to stop doing this; and next Sunday buy a paper which doesn't practise this.

#78 Pet care

As a planet, we bloody love our pets; an estimated 57 per cent of people in the world have one.[30] But pets, like humans, rely on a lot of plastic items – so how can we cut that back? Here are a few tips: buy stainless steel bowls for food and water; get an enamel cat litter tray; invest in a little shovel to pick up dog poo or buy biodegradable bags – and add to compost if you can; get a wooden hutch over a plastic cage; and give making your own dog food a go – you'll be surprised how many books there are on the subject!

#79 Pet toys

Look out for toys made from natural rubber, rope, canvas and other such materials. If you can't find any in your local pet shop then one such company you can buy from online is US-based Petfection, which prides itself on making all sorts of eco-friendly pet products. Several companies also make their toys from recycled materials, such as Cycle Dog, which make their balls from recycled bike tyres.

#80 Plastic-free kids' party

For the food – make everything hand-to-mouth friendly so you don't need to use plastic cutlery. Kids prefer this anyway. Avoid sweets and crisps that come in plastic packaging – instead make cakes/biscuits or support your local bakery. For the decorations – don't buy a plastic banner or badge. Make your own and get your child(ren) to help ahead of time as a fun activity. If you're strapped for time buy a card banner that hangs on string – don't be fooled by anything glittery or shiny unless it specifically says it's not plastic. For the party bag – either get paper ones or buy cheap plain fabric tote bags and get the kids to decorate them themselves with fabric pens. Fill with the types of tat your kids like (ideally of a non-plastic variety).

#81 A plastic-free Christmas

You don't need plastic to make this time of year magical, we have faith in you! Remember our present (pages 74 and 75) and wrapping (pages 77-79) tips and try to make other plastic-free decisions. Most advent calendars have a plastic tray insert keeping the chocolate in place – skip this by making or buying a fabric advent calendar and inserting your own gifts or (non-plastic wrapped) chocolates or sweets. If you have a plastic tree already, obviously don't chuck it out, but if you're thinking about buying one, don't. If you're desperate for a tree buy a real one and plant it in a pot in the garden after Christmas is done; it may start to brown but it should last for a good few years. Avoid buying Christmas lights – instead go for candles (decorating holders and tea-light jars makes a good pre-Christmas activity. Not suitable for the tree!) and swap tinsel and plastic baubles for wooden decorations; and plastic wreaths for real ones.

#82 Envelopes

Seriously, why do the envelopes with the little plastic windows still exist? We're all capable of writing the address out again (and who can fold the letter properly first time to get the typed address to show through anyway?). Also steer clear of envelopes lined with bubble wrap. Instead buy biodegradable envelopes or envelopes made from recycled materials.

#83 Ban the ballpoint

This one is a toughie. Some of us already use our tablets/laptops/phones to write everything with, but there are certain times when you just can't get around needing to scribble something down. And no one ever said 'the computer keyboard is mightier than the sword', after all. It's unlikely we can all return to feather quill and ink or use recyclable pencils the whole time, but we can revive the fountain pen. Most fountain pens are made from stainless steel and include a plastic nib, and will also take disposable plastic ink cartridges. But this will still amount to less plastic waste than the disposable plastic pens we pick up in handfuls from the office stationery cupboard and casually mislay in meeting rooms. Plus it makes a good personalised present for relations when you've run out of other ideas.

#84 Green gardening

Having green fingers can come with a plastic footprint: multi-purpose compost and fertilisers often come packaged in plastic, and plants in plastic plant pots. Growing plants and flowers from seed can help to get around this as you can get them in paper envelopes, and some plants also come in compostable containers – give your local garden centre a call to find out if they use them. If you can't get around buying plants in plastic pots, ask the shop if they'd like you to return them so they can reuse them afterwards. For soil, find a wholesaler near you who can deliver to your door, or you can try compost blocks such as Traidcraft's Fair Trade Coir Compost Block – you can get nine litres (2 gallons) of compost from one 650g pack, and the packaging is compostable.

#85 Hang on to your head phones

Your earbuds should be your best buds — look after them and you won't need several of them. Ever noticed that when you get a new phone the company often sweetly includes a new set for you? Decline these if your old ones are in good nick. Don't be tempted to cheat on your headphones on plane journeys either — reject the plastic wrapped free set in favour of your old faithfuls. There are also eco-friendly headphones available which utilise recycled plastics and natural materials: House of Marley and Lstn Sound are two of the brands you could try.

#86 Recycle your phone

Did you know that up to 80 per cent of your phone can be recycled?[31] Lots of phone companies will also give you some money back if you return your old phone to them, so make sure you take it with you when upgrading to a new one (or be a kind soul and hand it into a charity shop for them to reap the rewards). Often the phone companies will then recycle the parts of your old phone to repair others – great news!

#87 Plastic-free festival

This one takes a tiny bit more planning than you'd probably usually do – BUT it can save you money which you can spend on food and drink while you're enjoying yourself/make you feel less bad about the amount you spent on the ticket. First, don't be tempted by the two-litre bottles of water in the supermarket (heavy to carry anyway) – there will be a tap near you. Take a few reusable bottles and make use of it. If you like juice in the morning, take this in bottles too instead of the mini cartons with straws. Make your own snacks (that won't melt) rather than buying cereal bars and crisps. When buying drinks and food at the festival, be the trend-setter who takes their own glass/cup/plate/cutlery – the seller will thank you for not using up their stock and you will feel smug not adding to the overflowing bins. If you're dressing up, take biodegradable glitter (page 86) and other plastic-free touches. If you need a waterproof, don't forget that many of these are also made from a type of plastic – waxed jackets are a good way of avoiding this (and you can re-wax them to keep them waterproof as the years wane), but if they are too pricey for you then buy a second-hand waterproof from a charity shop. Also try and source plastic-free sunscreen (page 90). If you're investing in a tent, check the material – more than you think are made from plastic or other synthetic materials, but you can find ones made from cotton.

#88 Love your 3D glasses

It's all about 3D these days isn't it — so make sure you hold on to and look after your own set of 3D glasses; don't pick up a new pair every time you go to the cinema.

#89 Be mindful

Look after your things! It's as simple as that. Look after your phone; look after your headphones; look after your hair ties; look after your stationery; look after every item you own that contains plastic. The better care you give them, the less you will need to replace them and the less plastic that ultimately ends up in landfill or the sea.

SAVING

***

THE

***

WORLD

Our planet and its oceans need saving from other substances and elements aside from plastic, and it also needs saving from producing so much stuff – so here are some extra ideas on how to save the world.

#90 B. Y. O. C

Step away from the disposable chopsticks. 130 million pairs of chopsticks are made every day – EVERY DAY.[32] That's a lot of trees, deforestation and energy. So next time you're in a noodle house that doesn't do reusable chopsticks, don't be worried you'll look lame for using a fork. Forks are cool. Or be even cooler and carry a pair of reusable chopsticks in your pocket for just such an occasion.

#91 Long live the hankie

Remember the days when everyone had a handkerchief with their initials embroidered on them? Us neither, but it seems to happen quite a lot in old films. We say bring them back! In the US alone, upwards of 255 billion tissues are used each year.[33] That's a lot of resource and energy going into a disposable product. Plus, to bring it back to plastic for a moment, mini packs of tissues tend to come in a plastic packet, and boxes of tissues sometimes have a plastic lining. Hankie revolution here we come!

#92 Don't take the napkins

Just like with tissues (see 'Long live the hankie', page 109), a lot of resource and energy goes into the manufacture of napkins, and yet they seem to be given away and dropped willy-nilly. So long as you are a clean eater and not munching on a particularly mustard-y hot dog, you can probably do without one. Next time the nice cashier goes to hand you one with your coffee or morning pastry, politely decline — it's what the world wants.

#93 Recycle 101

This is one we should all be clued up on, but if we're all totally honest, few of us are 100 per cent sure what can and what can't be recycled or how each item should be prepared beforehand. The only way to find out is to check the guidelines on your local rubbish collecting service's website – lots give very helpful advice. Ideally you will be able to recycle plastic bottles and packaging, cardboard, paper, glass, metal packaging, food waste and garden waste. Recycling plants may struggle to get through it all so some of these products may end up in landfill anyway, but best foot forward and all that. If your local service does *not* provide the above, find out why not. If you can rally your community to make sure it is provided, what a hero you will be. Also remember to empty out any food or drink from the containers you are recycling, to rinse them and also to flatten any boxes or bottles. Remember that 'contaminated' items also can't be recycled, for example greasy pizza boxes (although any ungreasy bits can be).

#94 Go digital

There's nothing wrong with ye olde pastime activities: arts and crafts, a spot of gardening, running around in the great outdoors etc. But for some things – such as writing lists, reading articles, sending happy greetings (see page 76) etc. – going digital can save on resources. Where you can, put your phone and your tablet to use; you paid good money for them after all.

#95 Reuse paper

If you're a list maker or your kids like scribbling/ drawing pictures or you'd quite like to ignore the 'go digital' advice on page 112, then make use of any and all spare bits of paper you find. Start a 'paper drawer' if you will – squirrel away any one-sided junk mail (and envelope) you get; anything (non-private) you print out at work that has a blank page in the middle; the back of a programme from that local play you suffered through etc. and next time you need a bit of paper, you can make use of the paper already out there in the world rather than using up a brand new piece.

#96 Request less

When you're ordering anything online, follow up with the seller afterwards and ask them if they're able to send you the item with as minimal packaging as possible. Lots of retailers and individuals are happy to accommodate the request – it's ultimately less outlay for them. If you do receive something that comes in ridiculous packaging – a box in a box in a box in a box, for example – then let the retailer know they don't need to do that (or if you want to be a bit more passive aggressive about it, you can tweet a picture to social media – tends to make the retailer in question feel more accountable). Ask for the change you want to see, and all that.

#97 Plog, plog, plog

Scandinavians are such trend-setters and here they go again with 'plogging'. Plogging is picking up bits of litter, including plastic, whilst out on a jog. Any litter that doesn't get to a bin or landfill tends to find its way towards a drain or a river, which means it could ultimately get into the sea. And as we know, there's a plastic country three times the size of France floating in the Pacific we don't want to make any bigger. So next time you're out for a little run or a walk and you see a plastic bottle or napkin, be the lovely person who pops it in the relevant garbage vessel.

#98 Watch what you flush

The only things it is safe to flush down the toilet is (human) poo, wee and toilet paper (depending on which country you live in). Sewers often can't deal with or filter out other substances or toxins that come from anything else, and those products which should have gone in the bin can end up causing flooding or littering beaches.[34] Put another way – don't flush condoms, cotton pads, sanitary products including tampons, wet wipes or wipes of any kind *unless* the water board in your region has said so. Some products claim to be flushable; that's not always the case. Some people also argue that you can flush animal poo down the toilet; unless your local water board specifically says you can, assume that you can't.

#99 Watch how often you flush

When you flush the toilet, you use between 6 and 13 litres of water (1.6–3.6 gallons).[35] That's a lot of water and energy used each time when you take into account the process that happens after you flush and what measures are taken to make it clean again before it comes back into your cistern. And – fun fact – you really don't need to flush every time you do a wee – wee is not far off from sterile and it's flushing that causes the splashback of any germs.[36] So next time you pee, let it sit in the pot. This will also have the added advantage of making it really obvious when anyone's done a poo, so then you know to avoid the bathroom.

#100 Buy less and eat up

So many people these day seem to be foodies; you can't scroll down your Instagram homepage without seeing four smashed avocados and an artistically arranged waffle/pasta dish/baba ganoush. So why oh why does so much food seem to end up in the bin? The UK, for example, wastes 7.1 million tonnes of food in a year – environmentally speaking, stopping this would be the equivalent of taking 25 per cent of cars off the road.[37] So what can we do? A lot of initiatives, such as the Real Junk Food Project, have been set up in recent years to put to good use edible food supermarkets would otherwise throw out (because they are past their 'best before' date or the packaging is not seasonably accurate) – they collect and redistribute the food to cafes and school partnerships. There's also the app Olio, which allows business and individuals to post pictures advertising their leftover food so neighbours can come and take it off their hands. Aside from supporting (or starting?!) initiatives such as these in your own locale, you can do your bit by only buying the food you need and using the whole of it up. Freeze as much as you can to prevent it from going off before you've had a chance to eat it, and don't forget that – while you need to pay attention to 'use by' dates – the 'best before' date is just a guideline.

#101 Think before you use

Life is for living. Don't sit in the cold or go thirsty or be bored because you want to save on heating or you're worried about running the tap or turning on the TV. But DO think about how much energy you're expending. If it's winter and you're a bit cold but you're sitting in a T-shirt – before you stick on the heating, put on a sweater. Don't boil a full kettle if you're only having one cup of tea. Don't tumble dry your clothes if you can get them dry by hanging them in- or outside. Don't run the shower for five minutes before you get in it or let the tap run while you're brushing your teeth. Ultimately, just use your valuable common sense to only expend the energy you need to expend. Always remember that energy has a monetary value – the less you use, the less you spend.

REFERENCES

1 'Single-use plastic bag facts', www.biologicaldiversity.org (accessed 23/05/2018).

2 'Eight Million Tonnes of Plastic are Going Into The Ocean Each Year', www.iflscience.com (accessed 23/05/2018).

3 ''Turn the tide on plastic' urges UN', news.un.org (accessed 23/05/2018).

4 Simon Levey, 'Health of seabirds threatened as 90 per cent swallow plastic', www.imperial.ac.uk (accessed 23/05/2018).

5 Susan Smillie, 'From sea to plate: how plastic got into our fish', www.theguardian.com (accessed 23/05/2018).

6 'Hard Plastics decompose in oceans, releasing endocrine disruptor BPA', www.acs.org (accessed 23/05/2018).

7 A great initiative from Surfers Against Sewage – find out more at www.sas.org.uk/campaign/return-to-offender/

8 Sandra Laville and Matthew Taylor, 'A million bottles a minute', www.theguardian.com (accessed 25/05/05).

9 UK Tea & Infusions Association, 'Tea Glossary and FAQs' www.tea.co.uk (accessed 23/05/2018).

10 In London, according to iZetter. As reported by Isabella A, 'Londoners spend almost £2,500 a year on lunch', www.timeout.com (accessed 23/05/2018).

11 'What's a juice carton made from', www.revolve-uk.com (accessed 23/05/2018).

12 Sarah Knapton, 'Milk floats and glass bottles make a comeback as shoppers shun plastic', www.telegraph.co.uk (accessed 23/05/0218).

13 'About chewing gum – chewing gum statistics', www.chewinggumfacts.com (accessed 23/05/2018).

14 The Aluminum Association, 'The Aluminum Can Advantage', www.aluminum.org (accessed 23/05/2015).

15 In the UK, according to Recycle Now. As reported by Rebecca Smithers, 'British households fail to recycled a 'staggering' 16m plastic bottles a day, www.theguardian.com (accessed 23/05/2018).

16 'What is activated charcoal and why is it used in filters',

science.howstuffworks.com (accessed 23/05/2018).

17 Dermatologists Dr. Joshua Zeichner and Dr. Ranella Hirsch talking to Rachel Wilkerson Miller, 'How Often You Really Need to Shower (According to Science)', www.buzzfeed.com (accessed 23/05/2018).

18 Stats for the UK from www.whatprice.co.uk (accessed on 21/05/2018).

19 Bic, 'History', www.flickyourbic.ca/history (accessed 16/05/2018).

20 Jessica Dillinger, '7 Countries that spend most on toys', www.worldatlas.com (accessed 16/05/2018).

21 James Rodger, 'Revealed: The shocking amount parents are spending on children's toys per year', www.coventrytelegraph.net (accessed 16/05/2018).

22 In the UK each year. Global Action Plan, 'What can you do with unwanted presents', www.globalactionplan.org.uk (accessed 21/05/2018).

23 Poll conducted by GP Batteries, reported in Gemma Francis, 'Christmas: Brits will throw away 108m rolls of wrapping paper this year', www.independent.co.uk (accessed 21/05/2018).

24 Jo Confino, 'We buy a staggering amount of clothing, and most of it ends up in landfills', www.huffingtonpost.co.uk (accessed on 21/05/2018).

25 'What are the risk factors for lung cancer?', www.cdc.gov (accessed on 21/05/2018).

26 'Global Smoking Statistics', www.verywellmind.com (accessed on 21/05/2018).

27 Hannah Gould, 'Why cigarette butts threaten to stub out marine life', www.theguardian.com (accessed on 21/05/2018).

28 Liz Connor, 'Did you know women spend £18k on periods in their lifetime', www.standard.co.uk (accessed on 22/05/2018).

29 Will Coldwell, 'Hawaii becomes first US state to ban sunscreens harmful to coral reefs', www.theguardian.com (accessed on 22/05/2018).

30 'Most of world owns pets', www.petfoodindustry.com (accessed 22/05/2018).

31 'What to do with mobile phones', www.recyclenow.com (accessed on 22/05/2018).

32 Chris Davis, 'Saving trees one chopstick (level) at a time', www.chinadaily.com.cn (accessed 21/05/2018).

33 Linda Poppenheimer, 'Paper Facial Tissue – History and Environmental Impact', www.greengroundswell.com (accessed 23/05/2018).

34 'What to Flush', www.thinkbeforeyouflush.org (accessed 23/05/2018).

35 www.home-water-works.org (accessed 23/05/2018).

36 Rebecca Endicott, '8 Important Reasons you should always 'Let it Mellow' when you pee', www.littlethings.com.

37 www.lovefoodhate.waste.com (accessed 23/05/2018).